Properties of Matter
Describing and Measuring Matter

by Rebecca L. Johnson

Table of Contents

Millmark
EDUCATION

Everyone is here for a football game. Look at the photos.

What objects do you see?

I see _____.

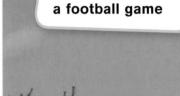

people watching a football game

What objects are smooth?

_____ are smooth.

What objects are rough?

_____ are rough.

What other ways can you describe the objects in the photos?

whistle

trombone

football

megaphone

popcorn

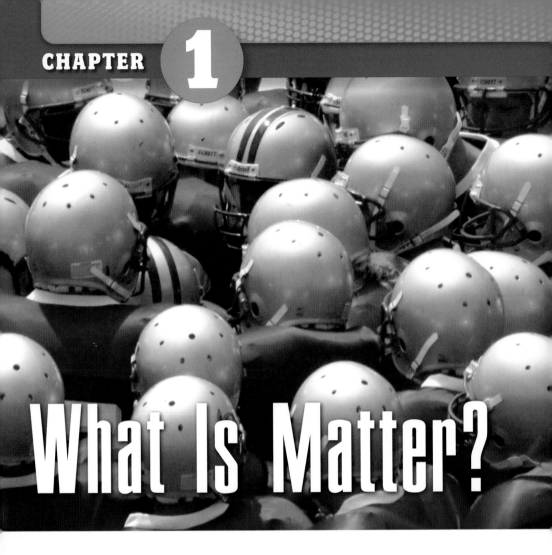

CHAPTER 1

What Is Matter?

All things are made of **matter**.
Matter is anything that takes up space.
A shoe is made of matter.

shoe

Matter is made of **atoms**.
An atom is a tiny unit that makes up every kind of matter.
Atoms are so small they cannot be seen.

binoculars

pom poms

bleachers

KEY IDEAS Matter is anything that takes up space. All things are made of matter.

Matter can be in different forms, or **states**.
A **solid** is one state of matter.
A solid has a **definite** shape and size.

An apple is a solid.
Bananas are solids, too.
A rock is a solid.
A tree is a solid.

A **liquid** is another state of matter.
A liquid has a definite size but no definite shape.
A liquid takes the shape of whatever it is in.

Water is a liquid.
Milk is a liquid.

pitcher

glass

milk

A **gas** is a third state of matter.
A gas has no definite shape or size.
A gas spreads out to fill all the space it can.

Air is a gas.

KEY IDEA Solid, liquid, and gas
are three states of matter.

▼ The air inside the
bubble is a gas.

bubble

Properties of Matter: Describing and Measuring Matter

INFER

Look at the photos.

What state of matter is the water in the goldfish bowl?

The water is _____ .

What would happen if you put the water in the aquarium?

The water's shape would _____ .

goldfish bowl

aquarium

MAKE CONNECTIONS

Liquids have a definite size. They do not have a definite shape. Find examples of liquids.

USE THE LANGUAGE OF SCIENCE

What is matter?

Matter is anything that takes up space.

CHAPTER 2

Describing Matter

feather

eggs

A bird's feather is matter.
The feather is light and soft.

A bird's eggs are matter.
The eggs are heavy and smooth.

Light, soft, heavy, and smooth are all **properties**.
Properties can describe what you see, touch,
taste, or smell.
Properties are used to describe matter.

wild turkey

SHARE IDEAS What properties would
you use to **describe** the wild turkey?

Remember that all matter takes up space.
Taking up space is a property of matter.

The skates take up space.
The car takes up more space.

States of matter are also properties.

The man's breath is a gas.
The honey is a liquid.
The apple is a solid.

breath

honey

apple

The apple's red color is a property.
The apple's round shape is a property.
The apple's sweet taste is a property, too.

These properties help describe the apple.

Explore Language

Adjectives
Red, **round**, and **ripe**
are adjectives. They
describe the apple.

KEY IDEA A property
of matter is something
you can see, touch,
taste, or smell.

COMMUNICATE

Look at the picture.
Tell a friend about
the statue's properties.

*The statue
looks _____ .*

*The statue might
feel _____ .*

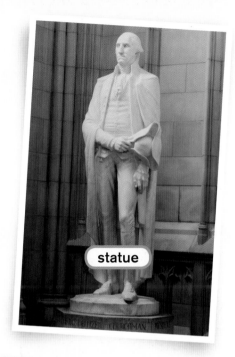

statue

MAKE CONNECTIONS

Find an object in the room.
Describe three of its properties.

 STRATEGY FOCUS

Make Inferences

Make inferences about properties of matter.
Name some properties of the car on page 12.
Tell how you know that the car has these properties.

glass marbles

Measuring Matter

Some properties of matter can be measured. When you **measure**, you find the size or the amount of something.

Mass is a property of matter that can be measured. Mass is the amount of matter in an object.

A **balance** can measure the mass of objects.
A **gram** is the basic unit for measuring mass.

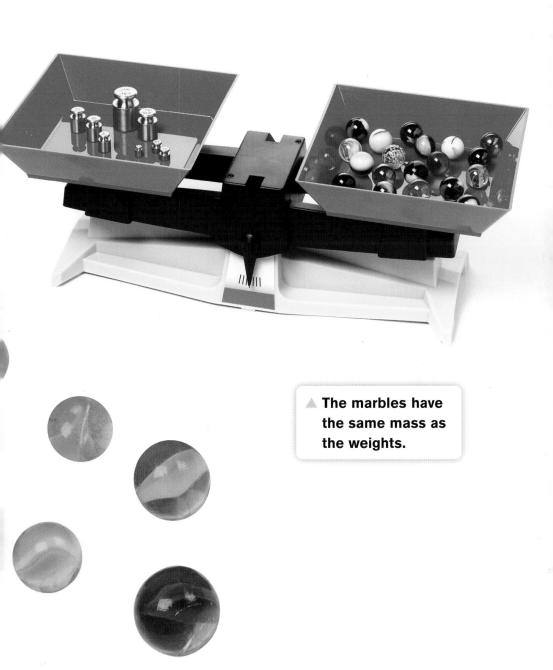

▲ The marbles have
the same mass as
the weights.

Volume is another property of matter that can be measured. Volume is how much space a kind of matter takes up. The **liter** is the basic unit for measuring volume.

Density is a third property of matter that can be measured. Density is how much mass there is in a certain volume of matter. Density is measured in grams per liter (grams/liter).

▲ The water in the bottle takes up 1 liter of space.

Measuring Matter	
What Is Being Measured	**Basic Unit of Measurement**
mass	gram
volume	liter
density	grams/liter

KEY IDEA Mass, volume, and density are properties of matter that can be measured.

▲ The density of water is greater than the density of oil. So oil floats on water.

INTERPRET DATA

	Sugar	Sand
mass	20 grams	40 grams
volume	1 liter	1 liter
density	20 grams/liter	40 grams/liter

Look at the table. Then interpret the data to answer the questions.

1. What is the basic unit used for measuring mass?

2. What is the density of the sugar?

The sugar's density is _____ grams/liter.

3. Which object has a greater density?

The _____ has a greater density.

MAKE CONNECTIONS

When you take a deep breath, air goes into your lungs. Tell how a balloon could show the volume of breath that comes out of your lungs.

EXPAND VOCABULARY

The following prefixes can be added to units, such as liters, to show smaller or larger amounts.

centi- = one hundredth (1/100)
kilo- = one thousand (1,000)

Which is larger, a centiliter of water or a kiloliter of water? Explain.

What Is an Atmospheric Scientist?

An atmospheric scientist studies properties of the atmosphere. The atmosphere is the blanket of gases that surrounds Earth.

Some atmospheric scientists study changes in the atmosphere that affect weather. Others study air pollution, such as smoke or harmful chemicals that enter the atmosphere.

Would you like to be an atmospheric scientist? Explain your answer.

▲ Atmospheric scientists study weather and air pollution.

◀ Atmospheric scientists use computers to study Earth's atmosphere.

Use Sensory Words to Compare

Properties can describe how an object looks, smells, tastes, or feels. Words that tell about information from your senses are called **sensory words**.

Sense	Examples of Sensory Words
sight	green, red, round, oval, big, small
smell	fresh, rotten, fragrant
taste	sweet, sour, bitter, salty
touch	soft, smooth, hard, light, heavy

With a friend, look at photo of the fruits on page 6. Use sensory words to describe the properties of the fruits.

Write a Comparison

Things can have different properties depending on their state. Choose one of the pairs below. Write a comparison about the objects. Tell whether each object is a solid or a liquid. Then explain how the properties in each state are the same or different.

- an apple / apple juice
- water in a bottle / an ice cube

Words You Can Use	
red	sweet
hard	smooth
clear	cloudy

Matter at the Market

BOTTLED WATER
two for
$1.00

each bottle 1 liter

Save $

JAR OF PEANUT BUTTER
$2.00

510 grams

Save $

Look closely at these coupons.

• What is volume of water in each bottle?

  The volume of water in each bottle is _____.

• What units are used to measure the peanut butter?

  The peanut butter is measured in _____.

Name some other foods that are measured in liters and grams.

Key Words

gas (gases) a state of matter with no definite shape or size

Air is a **gas**.

liquid (liquids) a state of matter with a definite size but no definite shape

A **liquid** takes the shape of its container.

mass (masses) the amount of matter in something

An object's **mass** can be measured.

matter anything that takes up space

You and everything around you are **matter**.

solid (solids) a state of matter that has a definite shape and size

The feather is a **solid**.

state (states) the form something is in

People drink water in its liquid **state**.

Index

MILLMARK EDUCATION CORPORATION
Ericka Markman, President and CEO; Karen Peratt, VP, Editorial Director; Lisa Bingen, VP, Marketing; David Willette, VP, Sales; Rachel L. Moir, Director, Operations and Production; Shelby Alinsky, Associate Editor; Mary Ann Mortellaro, Science Editor; Amy Sarver, Series Editor; Betsy Carpenter, Editor; Guadalupe Lopez, Writer; Kris Hanneman and Pictures Unlimited, Photo Research

PROGRAM AUTHORS
Mary Hawley; Program Author, Instructional Design
Kate Boehm Jerome; Program Author, Science

BOOK DESIGN Steve Curtis Design

CONTENT REVIEWER
Carla C. Johnson, EdD, University of Toledo, Toledo, OH

PROGRAM ADVISORS
Scott K. Baker, PhD, Pacific Institutes for Research, Eugene, OR
Carla C. Johnson, EdD, University of Toledo, Toledo, OH
Donna Ogle, EdD, National-Louis University, Chicago, IL
Betty Ansin Smallwood, PhD, Center for Applied Linguistics, Washington, DC
Gail Thompson, PhD, Claremont Graduate University, Claremont, CA
Emma Violand-Sánchez, EdD, Arlington Public Schools, Arlington, VA (retired)

TECHNOLOGY
Arleen Nakama, Project Manager
Audio CDs: Heartworks International, Inc.
CD-ROMs: Cannery Agency

PHOTO CREDITS Cover © Tipp Howell/Getty Images; 1 © Purestock/Getty Images; 2-3 © Tom Carter/PhotoEdit; 2a © O'Jay R. Barbee/Shutterstock; 2b, 8, 23a © Dennis MacDonald/age fotostock; 3a © Morgan Lane Photography/Shutterstock; 3b © Bob Daemmrich/The Image Works; 3c © Peter Ardito/Indexstock; 4a © Michelle Donahue Hillison/Shutterstock; 4b © Elnur/Shutterstock; 5a © Rafa Irusta/Shutterstock; 5b © Stockbyte/Getty Images; 5c © Stephen Finn/Shutterstock; 6 and 21 © Nicola Gavin/Shutterstock; 7 © Kord.com/age fotostock; 9a © Comstock/Punchstock; 9b © Central Aquatics; 9c and 9d Lloyd Wolf for Millmark Education; 10a and 23e © John Kaprielian/Photo Researchers, Inc.; 10b © Kenneth Thomas/Photo Researchers, Inc.; 11 © Toni Angermayer/Photo Researchers, Inc.; 12b © maxstockphoto/Shutterstock; 12a © Simplestockshots/Punchstock; 13a, 14, 23d © Corbis/Punchstock; 13b © Iva Barmina/Shutterstock; 13c © Oleg Boldyrev/Alamy; 15 © MedioImages/Punchstock; 16-17 © Stockbyte/Punchstock; 17, 18a, 18b, 19, 23c Ken Cavanagh for Millmark Education; 20a © Digital Vision/Punchstock; 20b © Photodisc/Punchstock; 20c © AP Images/Andy Newman; 22a © Carsten Reisinger/Shutterstock; 22b © MaRoDee Photography/Alamy; 23b © Iva Barmina/Shutterstock; 24 © 3777190317/Shutterstock

Copyright © 2008 Millmark Education Corporation

Published by Millmark Education Corporation
7272 Wisconsin Avenue, Suite 300
Bethesda, MD 20814

ISBN-13: 978-1-4334-0056-8

Printed in the USA

10 9 8 7 6 5 4 3